Geography Zone: Landforms™

VALLEYS

Emma Carlson Berne

PowerKiDS
press.

New York

Published in 2008 by The Rosen Publishing Group, Inc.
29 East 21st Street, New York, NY 10010

First Edition

Editor: Joanne Randolph
Book Design: Julio Gil

Photo Credits: All images Shutterstock.com.

Library of Congress Cataloging-in-Publication Data

Berne, Emma Carlson.
  Valleys / Emma Carlson Berne. — 1st ed.
      p. cm. — (Geography zone–landforms)
  Includes index.
  ISBN 978-1-4042-4203-6 (lib. bdg.)
  1. Valleys—Juvenile literature. I. Title.
  GB561.B47 2008
  551.44'2—dc22

                                    2007031813

Manufactured in the United States of America

Contents

What Is a Valley?   4

Valleys and Earth   6

V Is for Valley   8

U Said It!   10

A Rift Between Us   12

Creating a Canyon   14

Valleys Are Homes   16

Death Valley   18

Life in Death Valley   20

Keeping Valleys Safe   22

Glossary   23

Index   24

Web Sites   24

You may not have known it, but you have likely seen a valley before. A valley is a low place in Earth. Valleys have hills or mountains around them. Valleys sometimes have rivers at their bottom.

The sides and bottom of a valley can be bare and rocky or covered in trees and grass. A valley can be a hot, dry desert. It can be a wet, green **jungle**, too. Some valleys are narrow, while others are thousands of feet (m) wide.

The deepest valley in the world is in Tibet. This valley is called the Yarlung Zangbo Valley. It is 16,650 feet (5,075 m) deep.

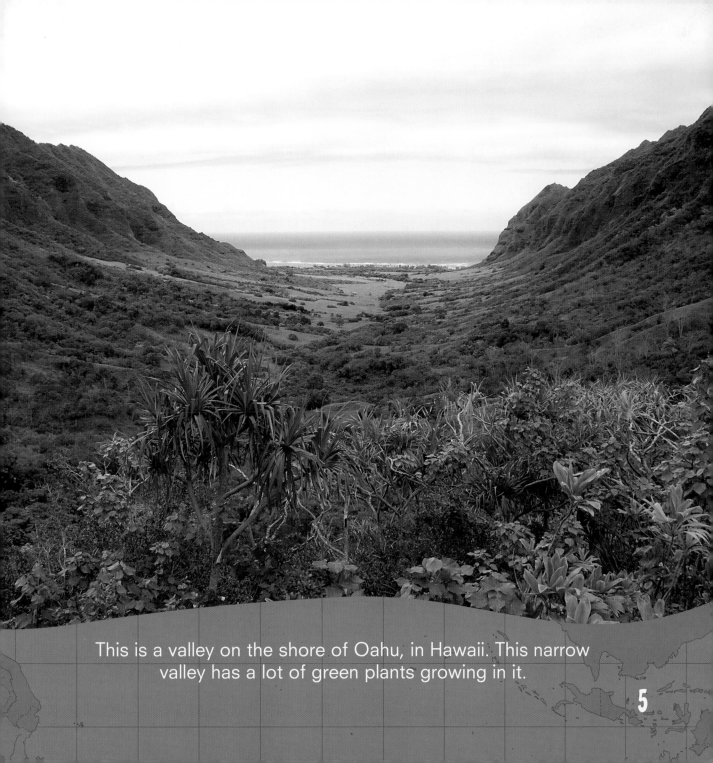

This is a valley on the shore of Oahu, in Hawaii. This narrow valley has a lot of green plants growing in it.

Valleys can be formed in a number of ways. Some valleys are formed when giant pieces of Earth slide around. These giant pieces of Earth are called **continental plates**. The continental plates sometimes crash into each other. When the plates crash together, they make mountains and valleys.

Sometimes the continental plates slide away from each other. They make a low place in Earth's surface when they slide away. The low place is a valley.

River waters can make valleys, too. Rivers that flow over the ground for a very long time can **erode** soil and rocks. The water makes a valley by washing the soil and rocks away.

Yosemite Valley, in California's Yosemite National Park, was created by glaciers. Glaciers are large masses of ice that move across the land.

Valleys have different shapes. Some valleys are in the shape of the letter *V*. These valleys are called V-shaped valleys.

A river makes a V-shaped valley. A river flows from its **source** in the mountaintops. The river carries rocks and sand in its **current**. The rocks and sand rub against the rocky mountain. Over time, the water and the rocks cut out a deep V shape. How steep and deep the valley becomes depends on how fast the river is moving. A slower river makes a gentle valley. A fast river makes a steep, deep valley.

The Engadine is a V-shaped valley in the Alps of Switzerland. Its name means "garden of the Inn River."

Some valleys are in the shape of the letter *U*. They are called U-shaped valleys. A U-shaped valley has a wider bottom than a V-shaped valley.

Sometimes the walls and bottom of a U-shaped valley are covered with grass and flowers. Scotland has lots of these valleys. The Scottish people call these valleys **glens**.

U-shaped valleys are made by huge **glaciers**. Many U-shaped valleys used to be V-shaped valleys. Millions of years ago, glaciers pushed into the V-shaped valleys. The glaciers carved out the sides and the bottom to make the U shape. The glaciers melted, but the valleys remain.

This is a U-shaped Scottish glen. Glens are commonly green and grassy.

Rift valleys are made when two continental plates pull away from each other. A **trough** appears where the plates had been touching. The trough is the rift valley.

Rift valleys are very big and deep. Sometimes, they fill up with water and become lakes or seas. Rift valleys often have **volcanoes** next to them.

When continental plates pull away from each other on the bottom of the ocean, a rift valley forms on the ocean floor. Underwater rift valleys can have volcanoes near them also.

Lake Nakuru is a lake in the Great Rift Valley, in Kenya, Africa. This lake is known for the huge groups of flamingos that come there.

Creating a Canyon

A canyon is a deep, V-shaped valley with very straight sides. The sides of a canyon are straighter than those of other V-shaped valleys.

Rivers make canyons. A fast, rough river flows through the bottom of a canyon. The river water washes over huge rocks. The water cuts the rocks away after many, many years. This makes the canyon.

The most famous canyon in the world is the Grand Canyon, in Arizona. The Grand Canyon is 277 miles (446 km) long. The Colorado River made the Grand Canyon. The Colorado River took six million years to cut the Grand Canyon.

The Grand Canyon is 6,000 feet (1,829 m) deep at its deepest point and its widest point is 15 miles (24 km) across.

Valleys Are Homes

Valleys are more than just landforms. They are homes! Many plants and animals live in valleys. The bottom of a valley usually has a river flowing through it. Willows, grasses, and other plants grow next to the river or in the water. Dragonflies like to live near rivers. Fish, turtles, ducks, and otters live in river valleys, too.

Deer, wild sheep, and wild goats live in valleys. They hide behind rocks and trees. They come to the river to drink at night. Wild cats, coyotes, and wolves hunt and eat the deer, goats, and sheep. They live in valleys, too.

Sheep eat grass in this valley in New Zealand. There are 45 million sheep in New Zealand.

Some valleys do not have a river at the bottom. These valleys are deserts. Death Valley is a big desert full of sand and rocks. It is in California. Death Valley is the hottest, driest place in North America. In the summer, the **temperature** in Death Valley is sometimes 130° F (54° C) during the day. On a winter night, the temperature can be 32° F (0° C), though. Brrr!

Rain falls only a few times a year in Death Valley. Even though it is dry, more than 1,000 different kinds of plants live there. These plants need only a little water to live. Foxes, sheep, and coyotes live in Death Valley, too.

Death Valley is home to the lowest place in North America.
It is also one of the hottest places on Earth.

Not many people can live in Death Valley because it is so hot and dry. A Native American **tribe** has lived in Death Valley for 1,000 years, though. This tribe is called the Timbisha Shoshone. The Timbisha Shoshone people are used to the hot temperatures.

Some people come to Death Valley just to visit. People like to see the beautiful colors of the rocks, sand, and plants. There is even a road through the valley called the Artist's Drive. It was made just for people who want to draw or paint in the desert.

This man has come to see the colorful rocks of the Artist's Palette, on Artist's Drive in Death Valley. Artists use palettes to mix paint colors.

Keeping Valleys Safe

People all over the world work and play in valleys. People grow crops on the valley bottom and fish in the river. Valleys are good ways to get around mountains. Therefore, people build train tracks in valleys. People also ski, boat, hike, and camp in valleys.

Many valleys are **polluted**. Trains puff smoke into the air. The dirty air gets trapped in the valleys. People have also dumped waste into the rivers, which hurts the plants and animals that live there. People are now learning how to live in valleys without hurting the plants and animals that live there.

Glossary

continental plates (kon-tuh-NEN-tul PLAYTS) The moving pieces of Earth's crust.

current (KUR-ent) Water that flows in one direction.

erode (ih-ROHD) To wear away slowly.

glaciers (GLAY-shurz) Large masses of ice that move down a mountain or along a valley.

glens (GLENZ) Small valleys usually covered with grass.

jungle (JUNG-gul) Land with lots of trees and plants, usually in a warm, wet place.

polluted (puh-LOOT-ed) Dirtied with harmful matter.

source (SORS) The place from which something starts.

temperature (TEM-pur-cher) How hot or cold something is.

tribe (TRYB) A group of people who share the same customs, language, and kin.

trough (TROF) A long, low place.

volcanoes (vol-KAY-nohz) Openings in the ground that sometimes shoot up hot, melted rock called lava.

Index

A

Artist's Drive, 20

C

Colorado River, 14
continental plates, 6,
    12
current, 8

D

desert(s), 4, 18, 20

G

glaciers, 10

glen(s), 10
Grand Canyon, 14

J

jungle, 4

R

rift valley(s), 12
river(s), 4, 6, 8, 14, 16,
    18, 22

S

Scotland, 10
source, 8

T

Timbisha Shoshone, 20
trough, 12

V

volcanoes, 12

Y

Yarlung Zangbo Valley,
    4

Web Sites

Due to the changing nature of Internet links, PowerKids Press has developed an online list of Web sites related to the subject of this book. This site is updated regularly. Please use this link to access the list:
www.powerkidslinks.com/gzone/valley/